DORSET 81

DORSET 81

Cape Dorset Graphics Annual

M. F. Feheley Publishers

5 Drumsnab Road, Toronto, Canada

ISBN: 0-919880-23-1 (Hardbound)
0-919880-24-X (Softbound)
ISSN: 0382-7747

Printed and bound in Canada

Photography/Terry Ryan, Jimmy Manning

Design/Howard Pain

Editorial/Sylvia Danter

Research/Patricia Ryan, S. M. Wright

Typesetting/Shervill-Dickson Limited

Film, Separation and Printing/Herzig Somerville Limited

This book was produced in Canada

Contents

Introduction

I can clearly recall my feelings when I was shown the first set of prints from Cape Dorset in 1959. I thought they were fakes simply because they were too good. How, after all, could a 'primitive' people have such a sophisticated sensibility? In addition to that they were superb draftsmen. It was difficult for me to accommodate my vision of the Eskimo to these beautiful works on paper. I began to question what the word 'primitive' really meant. I supposed it was a word used to compare other societies to ours, using our values rather than theirs. Using our standards, how was it possible for people who ate raw meat and lived in igloos to draw so beautifully?

Make no mistake, though these people reside in Canada they are immensely different. After all the years of our penetration into the North very few of them live in the South. Their preoccupations are so different from ours. The 1981 Cape Dorset print collection consists of fifty-two prints. Almost one third of them are concerned with the spirit world: the world of the *anghic* or shaman, legendary animals that one never sees, and animals that are becoming humans or humans that are becoming animals. Two thirds of the prints have to do with the animals that surround the people. Pudlo is about the only exception to this rule. In his prints we find imagery from the South: helicopters and aeroplanes. Strangely enough, however, these southern images have been converted into northern images. Having been captured and tamed, they are now genuine creatures of the North.

If you were to visit a northern community you would be confronted with many signs of a people in the midst of a transition that would make them an extension of the South. Why is it that so little of this shows up in their imagery?

Cape Dorset is the Florence of the North. Within the one small community, of few more than nine hundred people, there are separate groups of artists who are sculptors, lithographers and stonecut printers, artists known for their drawings, and artists recognized for their acrylic paintings on paper. A well-equipped printing-press shop has turned out a few modest but successful publications. Surely all of these various art forms must have had an effect on each other. Perhaps this is part of the reason why the Dorset scene is so successful. We know from our experience in the South that the metropolitan area can be a dynamic breeding ground because of the proximity of various creative groups which act and react upon one another. In the Dorset situation there are two separate audiences: the peer group in the North and the viewer in the South. It would be interesting to know how the peer group pressure works in Dorset.

There has been a continuous increase and spread of the audience for the Dorset prints: through Canada, into the rest of North America, and on into Europe. There is no other annual exhibition that excites such a large audience. All across Canada long lines form hours before the exhibitions open. Part of the reason, of course, is the investment factor. The prime reason, I believe, is the capacity of the Cape Dorset printmakers to communicate with their audience. That such a small community of artists should create such a large following is rather astonishing. We know, however, when it comes to the arts that even important single works can have amazing repercussions, and that it does not require quantity but quality.

Some contemporary artists in the South are seriously concerned with investigating the magic practices and the myths of the Indian and the Inuit. This is a reaction to our tendency to measure everything in empirical terms.

Perhaps we are growing weary of the new age of science. A peculiar inversion is taking place. The subject-matter that we commonly see in Dorset prints, that is, the subject of the anthropomorphic and the spirit, has become the patron saint of the new avant-garde.

It is well over twenty years since it all began and there has been no let-up in the intensity, vitality or surprises. How fortunate we are to have had a front seat to each year's output and to have been allowed the privilege of an intimate view into the heads of a foreign and exotic society.

Avrom Isaacs
June, 1981

A Part of My Life

When I became old enough to go around by myself, without my sister watching over me, I started hunting birds. I quietly studied their movements and grew to love them. Later I began to observe the varieties of birds that arrived in the arctic every spring. I studied the shapes, sizes, colours and sounds of different types of birds.

During the hunt, other animals didn't interest me as much as birds. As a young boy I hunted ptarmigan, sea pigeons, murres, ducks and geese for food for my family. On the way home I would study the dead bird's feathers, feet, shape, size and colouring. At that time I never knew that I was going to be drawing birds in the future.

Animals are much more difficult to observe than birds. There is also the added danger of approaching such animals as the polar bear, walrus, wolf or even caribou. Whenever I could, however, I enjoyed watching the powerful movements of the animals, and envied their gracefulness and intelligence.

When I was younger I often felt grief over killing a beautiful animal, even though it was for food. The older people wisely told us that killing animals must be purposeful, and not simply for sport. Long ago the old people used to tell us that animals were made by a great spirit in order for the Inuit to have food. I learned that a wise hunter should not boast about his success at the hunt, or he would have difficulty in future hunts.

Long ago the weapons used by the Inuit were harpoons, spears, and bows and arrows. Harpoons were used for hunting seal, walrus and whale, whereas spears were used for game such as the polar bear, caribou and rabbit. The hunter required skill and speed in order to achieve success at hunting. Today animals are hunted in a much different manner, using guns and faster methods of transportation.

In the future the time may come when the Inuit no longer hunt game. Therefore I record on paper these events from the spoken words of my people and from my imagination. I have come to the end of this small part of my life.

Kananginak Pootoogook ᑲᓇᒥᓇ

Translated by Pia Pootoogook

Edited by Patricia Ryan

Caribou, Dorset Area

Kananginak Pootoogook at Enukso Point, South Baffin Island

Kananginak's Comments on His Stonecuts

A Good Day Fishing (Page 28)
When lake-ice deteriorated in spring and holes developed in its surface, the Inuit would fish with spears and hooks. Seagulls always hovered overhead hoping to steal a fish.

I Was at the Rudder (Page 29)
Many years ago the only source of wood for the Inuit was what they could gather on the shoreline. The driftwood would be used in the construction of kayaks, umiaks and tent frames. The women sewed the sealskins together with sinew and then stretched them over a frame to make an umiak. Intestine skins, sewn together, functioned as sails.

Tiriak Battles the Owl (Page 30)
Many species of birds, some of which remain all year, are found in the arctic. Falcons, hawks and snow owls usually kill smaller animals for their food. The owl feeds mostly on the lemming and the weasel.

The Raven and the Goose (Page 31)
There is a legend about the Snow Goose and the Raven. The Raven desired the Snow Goose for a mate. She continually refused, however, because she knew that during the long migration south, over the open seas, the Raven would not be able to rest when weary. Nevertheless the Raven persisted and the Snow Goose agreed. During the long flight south the Raven frequently alighted on the back of the Snow Goose to rest. Eventually the Snow Goose tired of this and, while resting on the open water, turned on her side allowing the Raven to drown.

Sharks Near Our Boat (Page 32)
Bows and arrows, harpoons and spears were hunting tools used by the Inuit before they acquired guns. The Inuit had to hunt game in order to survive. Not only was the meat of the animal used, but also the skins which were made into clothing and other useful items.

Feeding on Flies (Page 33)
The migration of birds north makes the Inuit joyful. They know that spring has arrived. Years ago, in times of hardship, birds were a source of food for the Inuit. The snow bunting is one of the first birds to arrive in the arctic.

Mother Protects Her Young (Page 34)
Although the owl is known to be king of birds, the small Arctic tern is brave enough to oppose it. The owl, protecting her young with her wings, is unable to fight back and kill the tern. Terns, light and swift, use their sharp beaks as weapons.

Dance of Love (Page 35)
In the spring eider ducks migrate north to mate and lay their eggs. Mating begins in the middle of May. The male and female remain together until the eggs are hatched. They feed on clams and snails from the sea.

Sijariaq (Page 36)
The sandpiper arrives in the arctic after the lake-ice has melted. Before she lays her eggs, the sandpiper feeds continuously so that she is not required to leave her nest.

The Drawing and Painting Studio in Cape Dorset

Cape Dorset artists have traditionally done their drawing at home. Pitseolak prefers to sit up in bed while Kenojuak often kneels over her work on the floor. Many artists simply use the kitchen table. All use drawing materials distributed by the Co-op when finished works are brought in for purchase. Rag-paper, pre-cut to fit various printing surfaces, coloured pencils and black felt-tip pens are provided.

In 1976 some of the artists were given the option of working in two small studios. There they had the opportunity to experiment with acrylic paint left for them by K. M. Graham, a well-known Toronto painter who had paid a working visit to Cape Dorset. Over a period of two years Etidloie and Eliyakota expanded upon their drawing, while Kingmeata, Pudlo, Lucy, Napachie, Eegyvudluk and Soroseelutu worked in paint and coloured pencil on paper. Unfortunately after two years the studios were closed. A massive inventory of paintings on paper, high maintenance and heating costs, and the generally poor condition of the buildings prompted the closure.

The idea of a common studio was not however abandoned. Both the positive critical comment on the acrylic paintings and the desire of the artists to share working space kept the idea alive. In February of 1981 a larger and more efficient area, which had originally housed a jewellery project, was officially opened as the new drawing and painting studio.

Relatively new Dorset artists, including Haunak Mikigak, now benefit from the stimulus of working beside the more experienced artists. The proximity of the artists to the printshops has also served to enhance the collaboration between themselves and the printers.

The artists, whose works are illustrated here, are currently using the drawing studio. These include Kananginak Pootoogook, Etidloie Etidloie, Kingmeata Etidloie, Napachie Pootoogook, Eegyvudluk Ragee, Eliyakota Samuellie, Pudlo Pudlat, Haunak Mikigak, and Pitaloosie Saila.

W. Brannen

Eliyakota *Untitled 1981*
Coloured pencil, black felt pen
56.5 x 76.5 Signed by the artist

Haunak *Untitled 1981*
Acrylic paint, coloured pencil, black felt pen
56.5 x 76.2 Signed by the artist

Napachie *Untitled 1981*
Acrylic paint, coloured pencil, black felt pen
56.3 x 76.3 Signed by the artist

Eegyvudluk *Untitled 1981*
Acrylic paint, coloured pencil, black felt pen
56.5 x 76 Signed by the artist

Etidloie *Untitled 1981*
Coloured pencil, black felt pen
56.6 x 76.2 Signed by the artist

Kingmeata *Untitled 1981*
Acrylic paint, coloured pencil, black felt pen
56.5 x 76 Signed by the artist

Pudlo *Untitled 1981*
Coloured pencil, black felt pen
56.5 x 76.2 Signed by the artist

Pitaloosie *Untitled 1981*
Coloured pencil, black felt pen
56.5 x 76.5 Signed by the artist

Kananginak *Untitled 1981*
Coloured pencil, black felt pen
56.5 x 76.5 Signed by the artist

Cape Dorset Annual
Graphics Collection 1981

1 Aoudla Pudlat *Imaginary Bird*

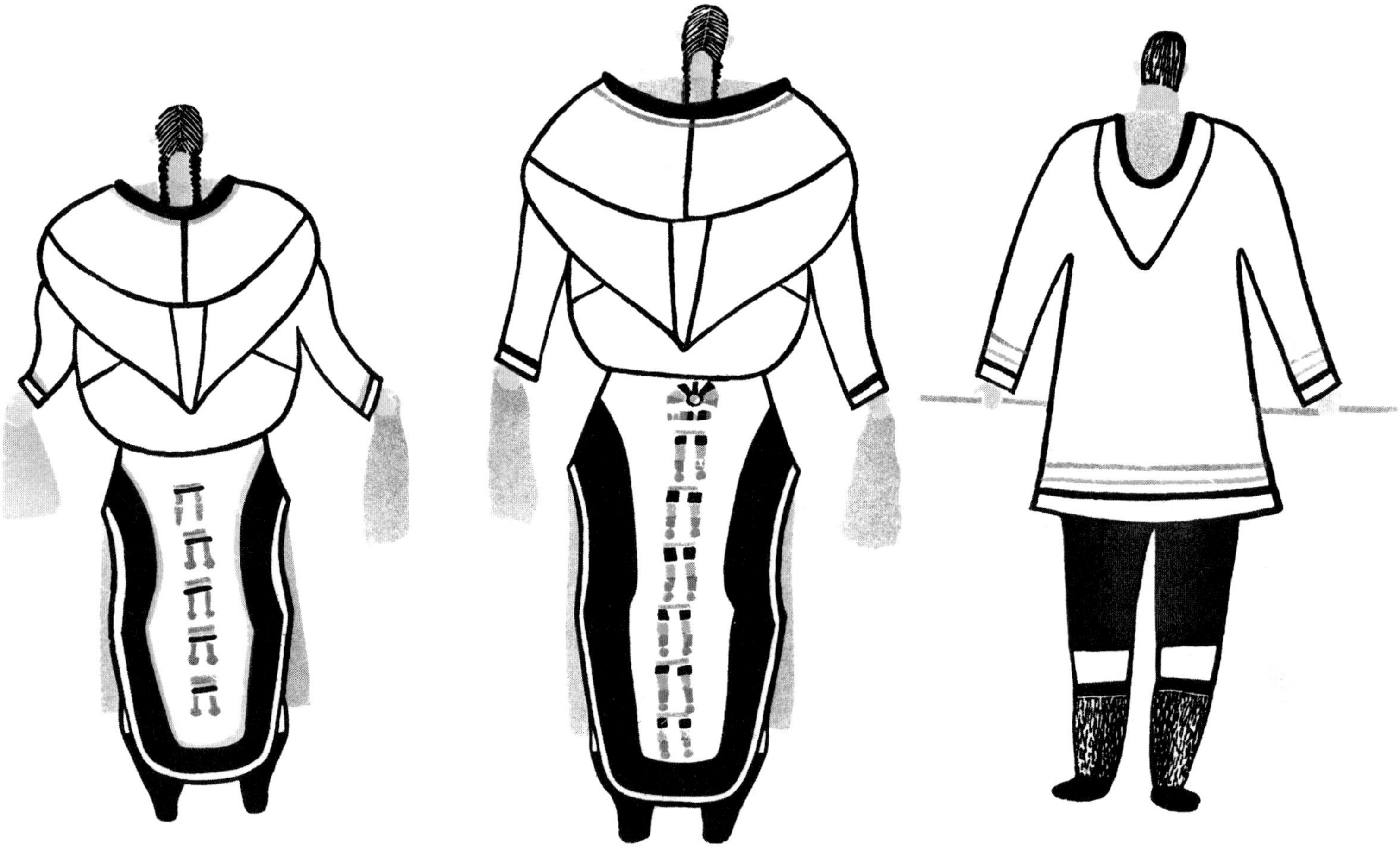

2 Echalook Goo *Inuit*

3 Eegyvudluk *Angako Quviasuktu*

4 Eegyvudluk *Ishuk*

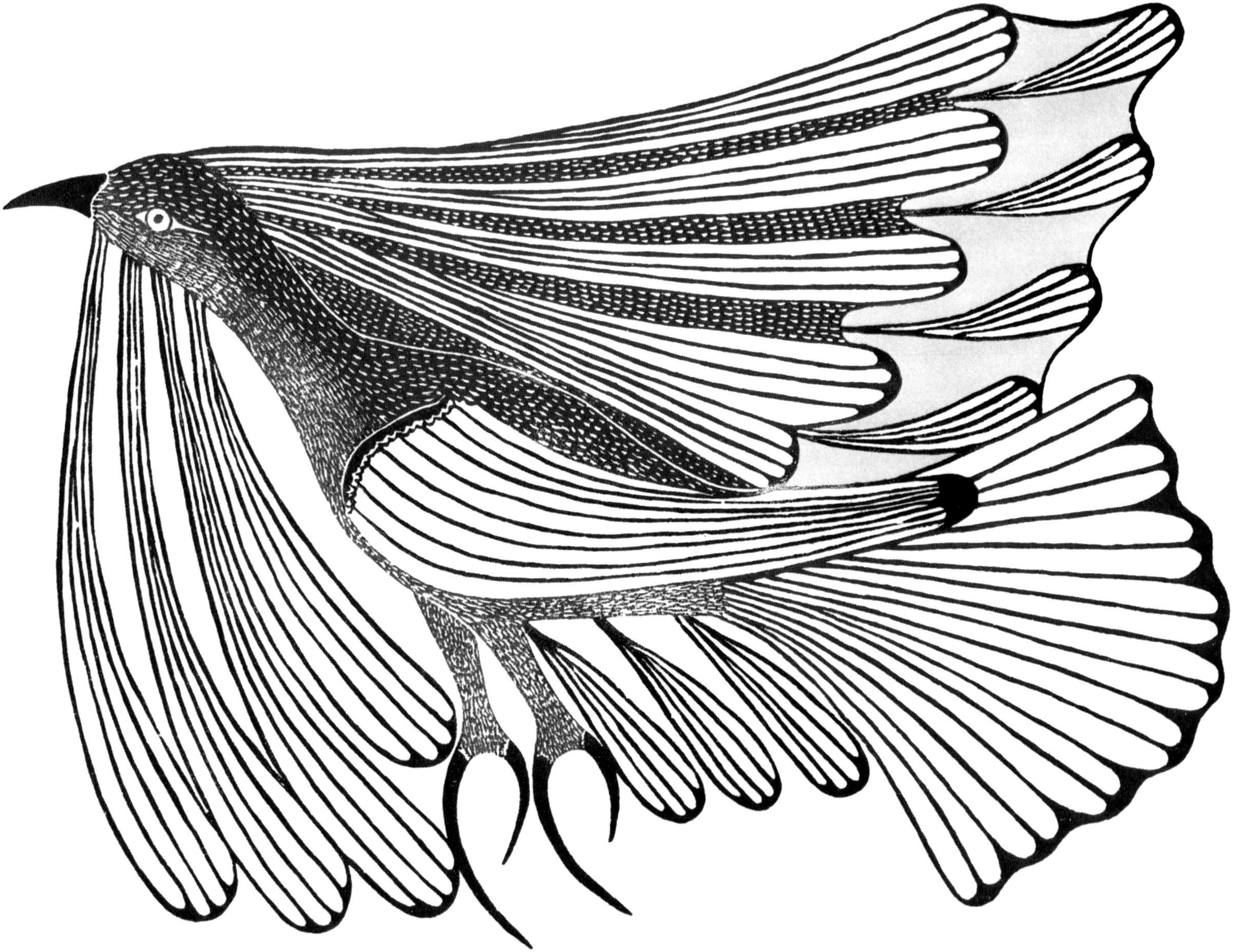

5 Eliyakota *Flying in a Strong Wind*

6 Etidloie *Marching Birds*

7 Etidloie *Birds Among Willow Trees*

8 Haunak *First Goose of the Spring*

9 Haunak *Owl Attacking Prey*

10 Jamasie *Sailing in a Beautiful Boat*

11 Kakulu *Disturbed While Nesting*

12 Kakulu *The Clothing I Make*

13 Kakulu *Family Members Return*

14 Kananginak *A Good Day Fishing*

15 Kananginak *I Was at the Rudder*

16 Kananginak *Tiriak Battles the Owl*

17 Kananginak *The Raven and the Goose*

18 Kananginak *Sharks Near Our Boats*

19 Kananginak *Feeding on Flies*

20 Kananginak *Mother Protects Her Young*

21 Kananginak *Dance of Love*

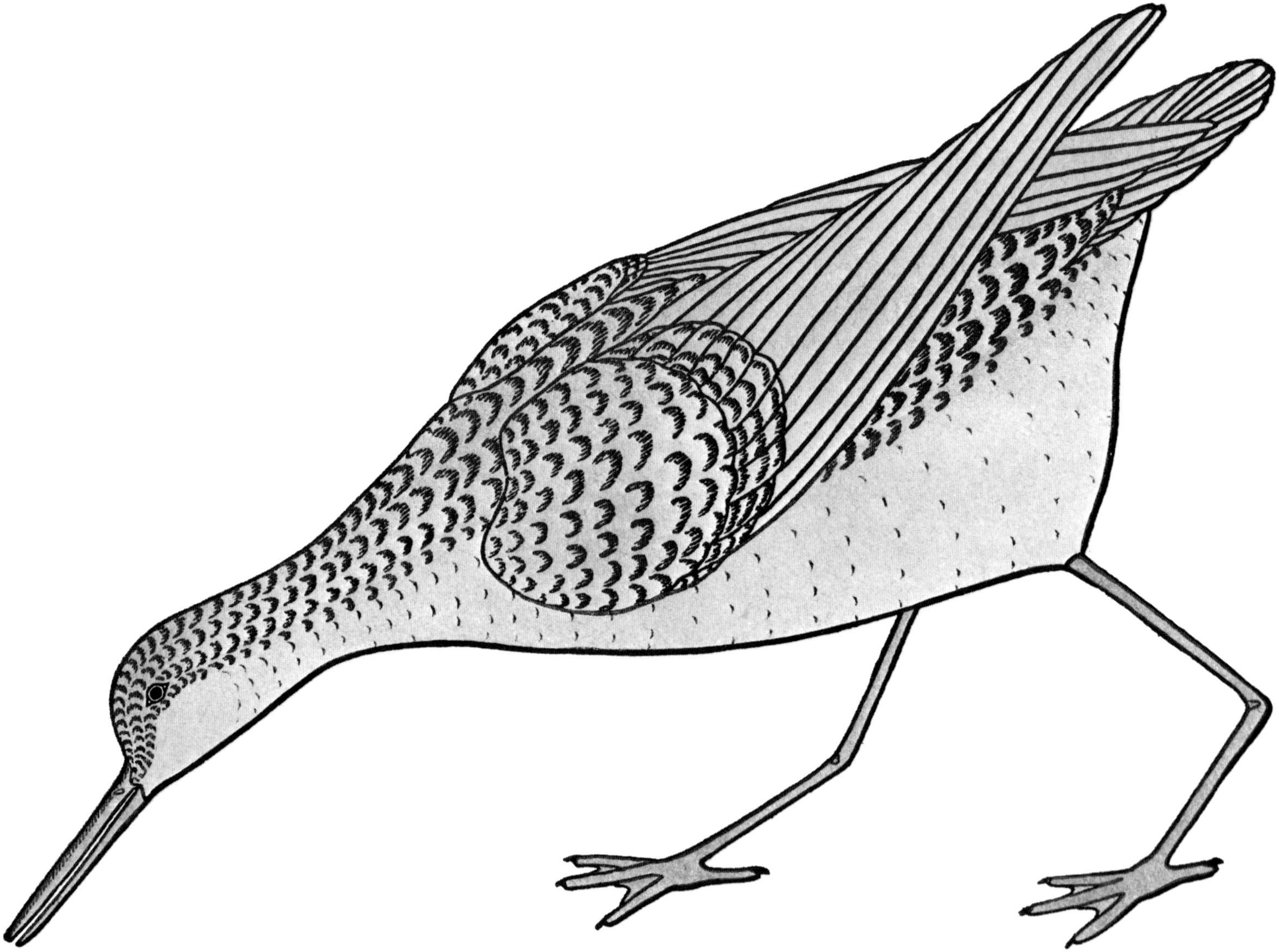

22 Kananginak *Sijariaq*

23 Keeleemeeoomee *Some Feed, Some Watch*

24 Keeleemeeoomee *We Follow the Caribou*

25 Keeleemeeoomee *Two Birds Walking*

26 Kenojuak *Watching My Animals*

27 Kenojuak *The Owl Rules the Summertime*

28 Kingmeata *Talelayou and Fish*

29 Kingmeata *Joyful Young Man*

30 Kingmeata *Hawk and Bear Converse*

31 Kingmeata *Large Bird in Sunlight*

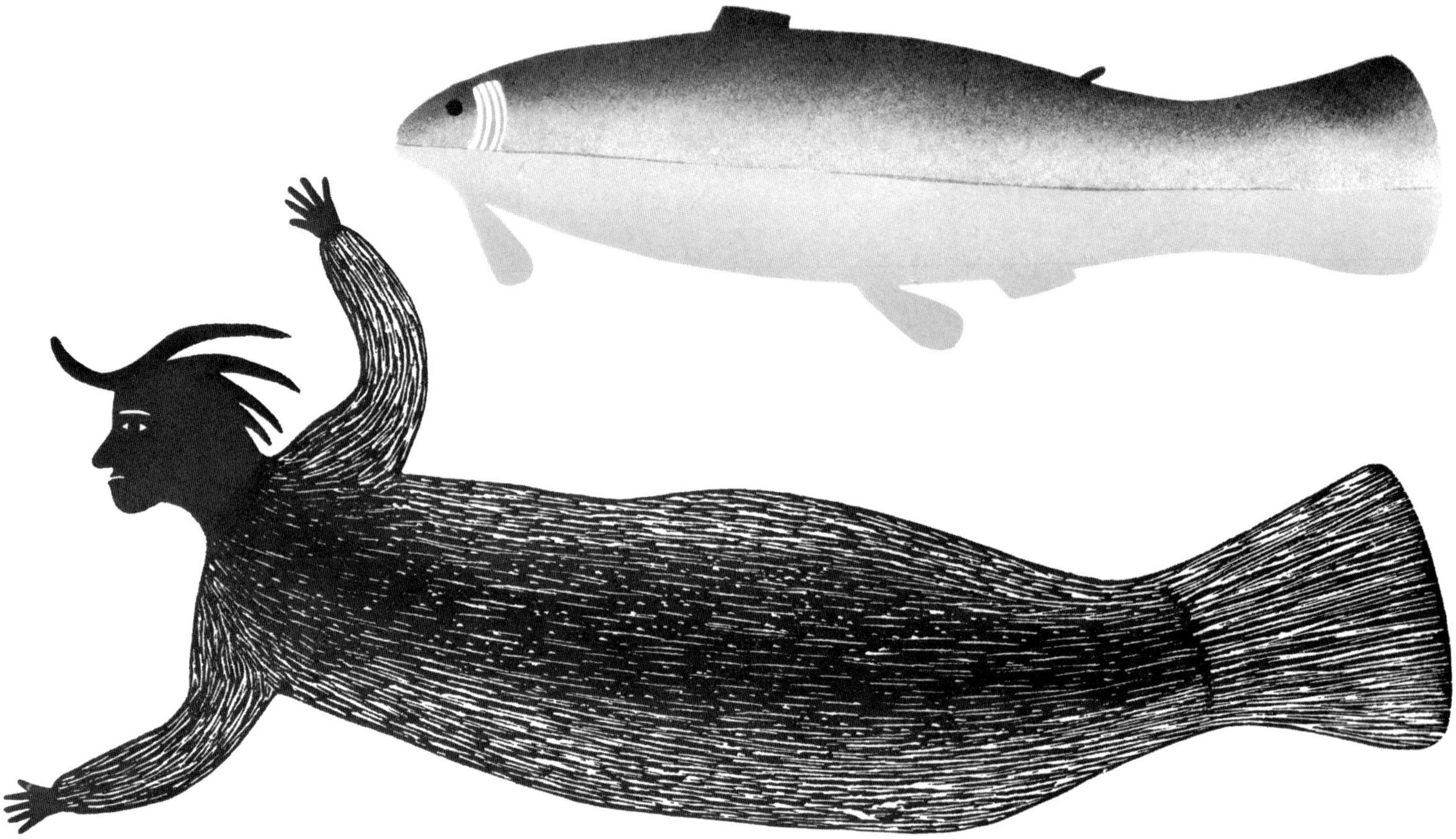

32 Kingmeata *Talelayou and Spirit Fish*

33 Kingmeata *Bird Sentinel*

34 Kudjuakjuk *Birds Eat Mosquitoes*

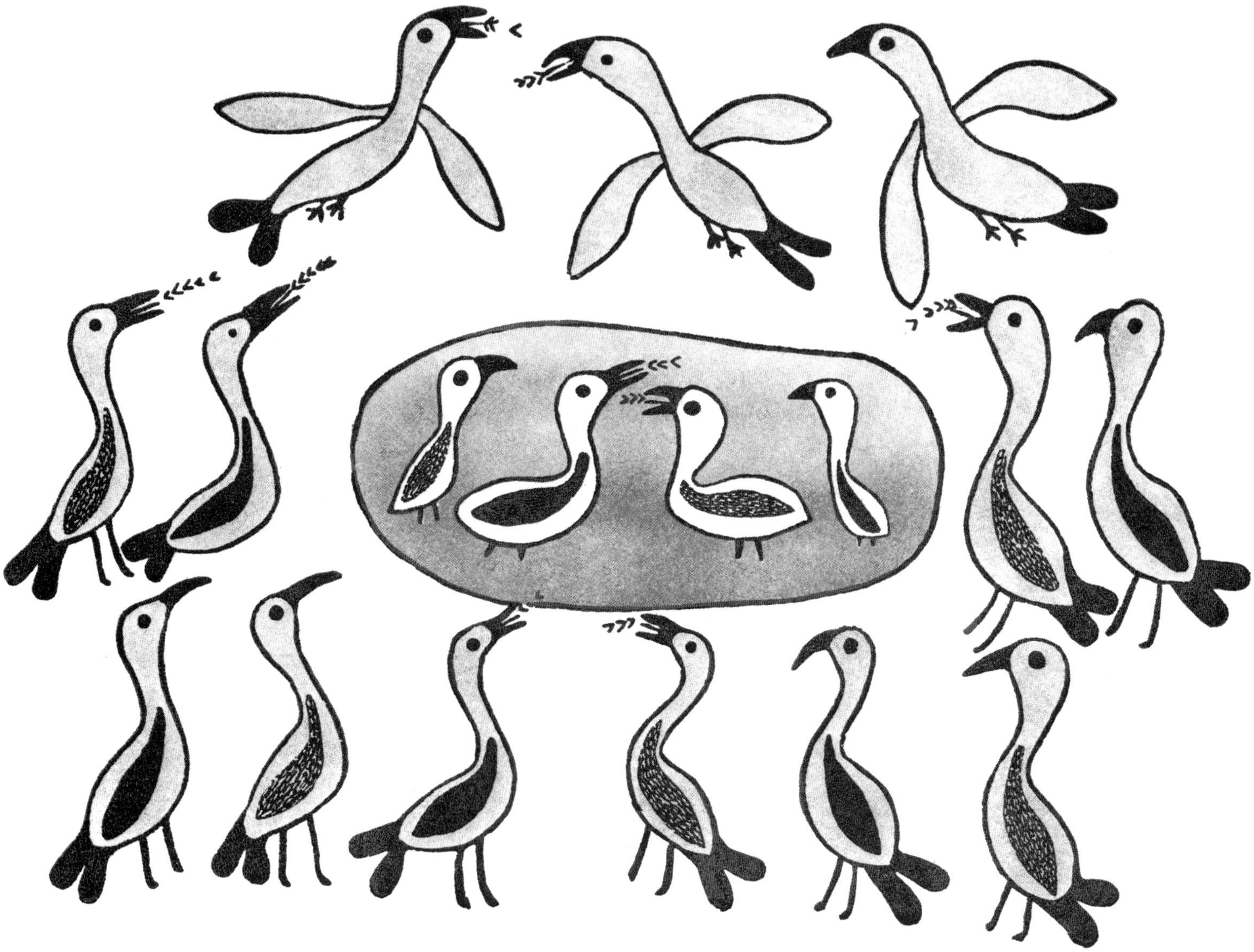

35 Kudjuakjuk *Seagulls on Spring Ice*

36 Lucy *Talelayou, Giant Birds and Men*

37 Lucy *Attacked by a Giant Bird*

38 Lucy *The Owl Looks On*

39 Lucy *A Bird Steals a Fisherman*

40 Lucy *Birds Fishing*

41 Lucy *Trio of Owls*

42 Lucy *Catching Moulting Birds*

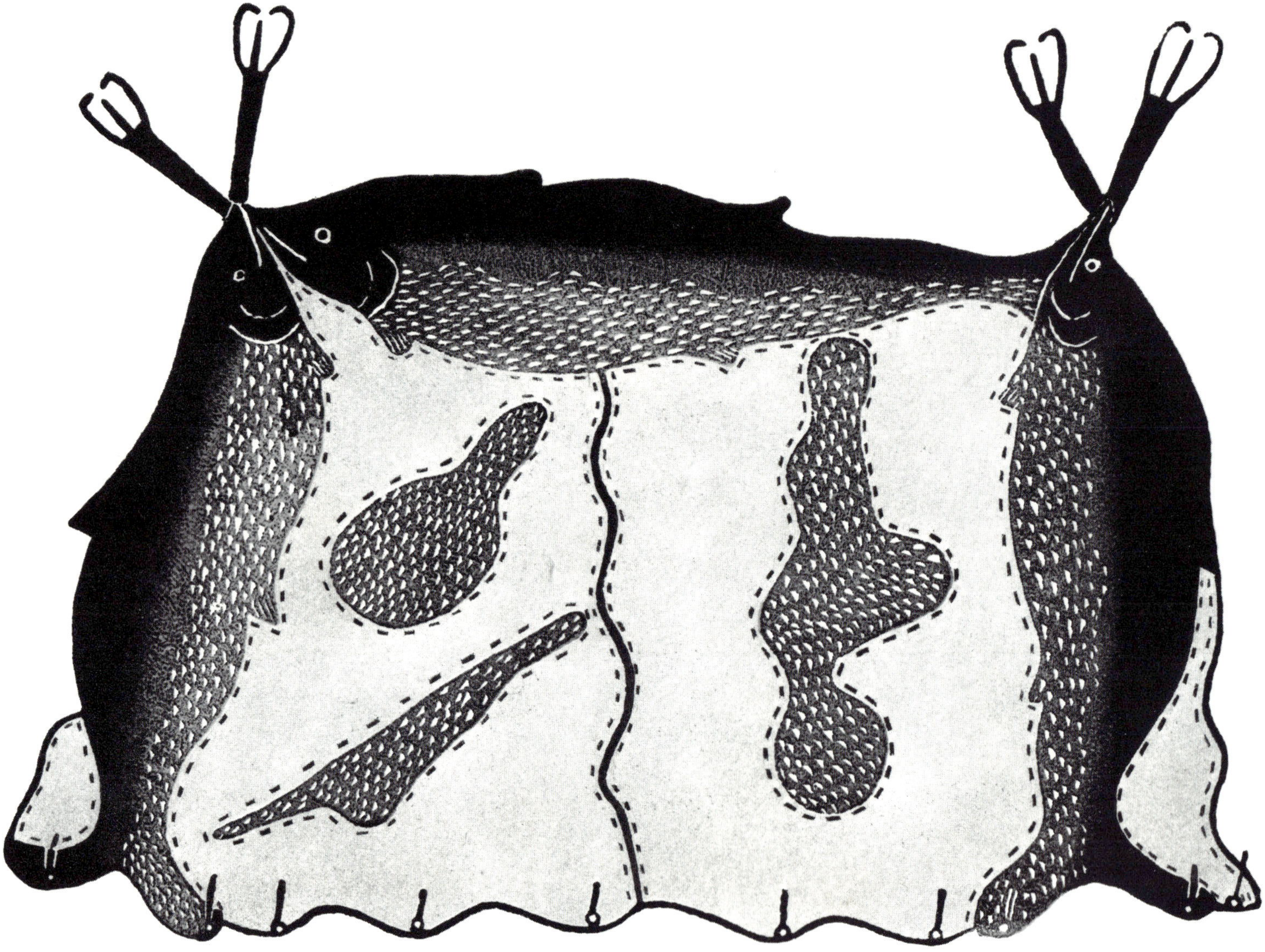

43 Napachie *A Tent at the Fish Lake*

44 Pitseolak *The Eggs Begin to Crack*

45 Pitseolak *Dogs Eat the Seal*

46 Pitseolak *Bird Escapes My Stone*

47 Pudlo *Ship of the Loon*

48 Pudlo *We Travel With the Loon*

49 Pudlo *Loon Speaks to the Walrus*

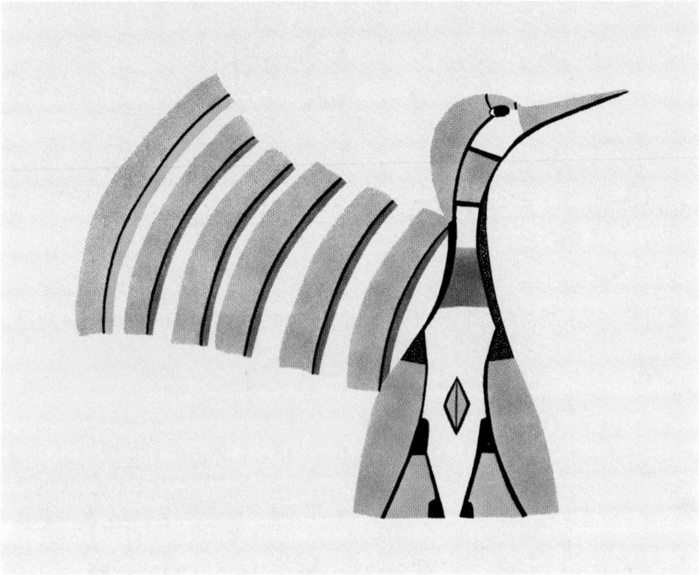

50 Pudlo *Ready to Fly*

51 Pudlo *The Loon and the Fish Speak*

52 Pudlo *Ptarmigan in Sunlight*

Lithography in Cape Dorset

Unlike stonecut printmaking, which was created by and for the Cape Dorset artists, hand lithography has been an artist's print medium since the late 1800's, and it is ironic that it is considered by most Inuit artists to be the newer, more modern process. In Cape Dorset, the lithographic process did not draw technicians from the existing ranks of older printers, but attracted instead a younger group of craftsmen that came together for this purpose early in the summer of 1974. It was at that time that the first large multi-coloured prints were produced. Due to inadequate studio facilities, the staff was unable to complete the usual edition of fifty prints.

The first serious experiment was undertaken in 1974 with a suite of prints taken from the drawings of the recently deceased Peter Pitseolak. This man, whose work differed greatly from that being produced by other Co-op artists, was an artist-photographer-historian who created art of a colourful boldness which would be reflected in the work of many in the years to come.

This often controversial use of colour remains an important concern of the Cape Dorset lithographers. The work of Pudlo Pudlat, the most regularly published of the Cape Dorset artists, demonstrates an increasing awareness of this medium as the perfect vehicle to promote his sophisticated use of colour in form. With the designs of Mayoreak, it becomes a finely-tuned instrument capable of expressing the exacting detail and colour for which her original works are noted. In the hands of Aoudla Pudlat, performing the dual function of lithographer and artist, this method promotes the most unusual approach in the studio since the editioning of Peter Pitseolak's designs seven years ago.

After years spent in developing a full range of skills in this medium, it is the master printer's role to interpret the original works of the artist. In Dorset two senior lithographers, Pitseolak Niviaqsi and Aoudla Pudlat, have become adept at dealing with the different requirements of the Co-op artists. They are aided in transferring the original sketches to plate and stone by Keatshuk Niviaqsi. These three have been training a third lithographer, Pootoogook Keatshuk, who has been with the Co-op since 1977 as a typographer and, more recently, as Aoudla Pudlat's press assistant.

In addition to the regular tasks of handling the product of their own studio, the lithography staff plays an active role in handling the published works of the stonecutting and etching workshops. Since 1976 they have also provided the support mechanism for those artists involved in experimenting with acrylic painting and the drawing mediums. Curator Liamiki Toonoo purchases these works, and also manages a small framing shop at the Co-op.

The studio has not often been worked to the limit, but the growing demand for commissioned works is rapidly filling the schedule. In spite of the frequency and efficiency of air transport, the lithographic medium can be a difficult one to manage and supply. In 1974 when a shipment of vinyl lacquer, essential to the lithographic process, was missing from an order, thirty-six containers of nail polish from the Hudson's Bay Company replaced the missing chemical. In 1981 we were not so fortunate. Plans for an annual show made up entirely of large multi-coloured prints were abandoned when it became evident that an order for aluminum plates would not be filled in time. The small stones, so useful in the early days, were taken from storage and Kananginak created a series of images to compliment these rare limestone surfaces. Literally every available surface was used. One such stone, necessary in the series, was printed with its irregularities showing through. "Building a Kayakkovik" depicts a vibrant sky against a crisp black drawing. The black of the stone has left the impression of an oval chip and a missing corner.

The following are the employees of the lithography studio in Cape Dorset:

Plate and stone preparation, Keatshuk Niviaqsi
Printer, Pitseolak Niviaqsi and assistant, Pabinak Petaulassie
Printer, Aoudla Pudlat and assistant, Simeonie Jamasie
Printer trainee, Pootoogook Keatshuk and assistant, Kavavow Pootoogook
Curator, Liamiki Toonoo

W. Brannen

L1 Aoudla Pudlat
Colourful Bird and Fish

L2 Aoudla Pudlat
Bird with Large Fish

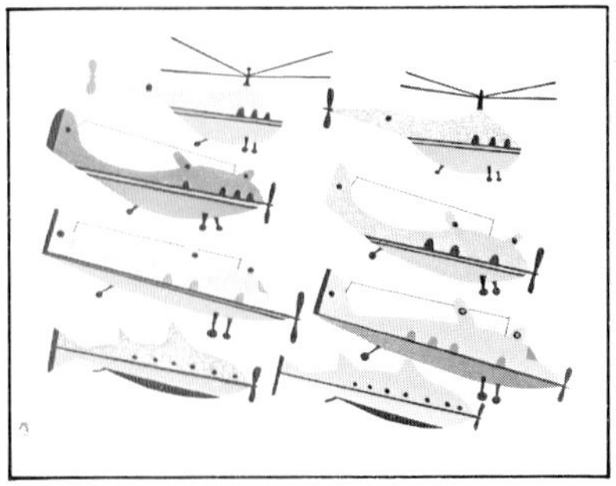

L3 Etidloie
Aircraft Becoming Sea Animals

L4 Haunak
Geese in Spring

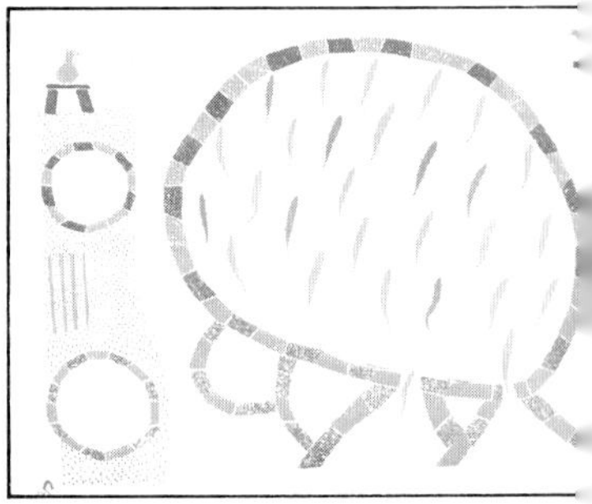

L5 Jamasie
Fish Weir

L6 Jamasie
Walrus and Friend

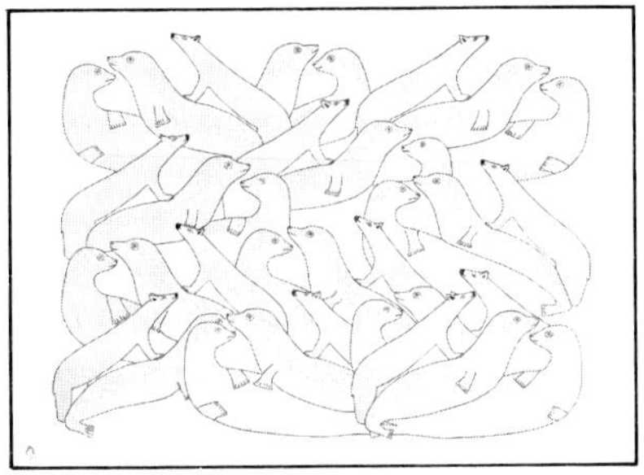

L7 Kakulu
The Animals Come Together

L8 Kananginak
Nirlik

L9 Kananginak
Kopinawa

L10 Kananginak
Swan

L11 Kananginak
Metik

L12 Kananginak
Tern Attacking Eider Duck

L13 Kananginak
Gulls at Sunrise

L14 Kananginak
Going into the Tent

L15 Kananginak
Building a Kayakkovik

L16 Kananginak
Hunter at Aglu

L17 Lucy
Young Owl in Sunlight

L18 Lucy
Boy at Home Alone

L19 Mayoreak
Putting Up the Tent

L20 Mayoreak
Cleaning Fish

L21 Pauta
Mother with Son

L22 Pitaloosie
Great Owl

L23 Pitaloosie
The Young Owl

L24 Pitaloosie
Dancing Ptarmigan

L25 Pitaloosie
Mother

L26 Pudlo
Muskox and Friends

L27 Pudlo
Large Loon and Landscape

L28 Pudlo
Boat and Airplane

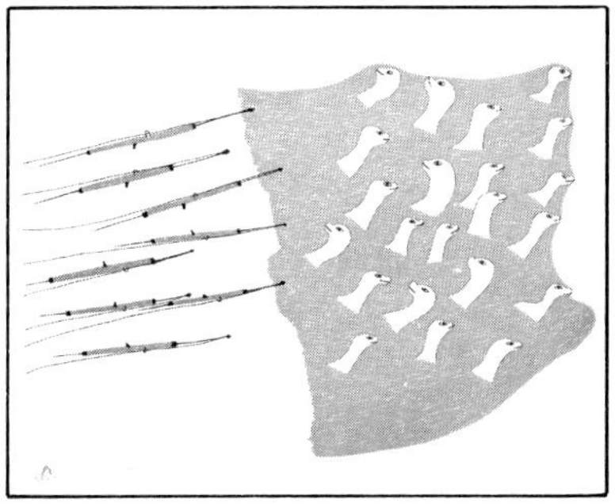

L29 Pudlo
Harpoons and Seals

The Mintmark Press, 1981

In 1981 The Mintmark Press, Toronto, Canada, commissioned a limited edition etching by Kenojuak Ashevak for inclusion in the book, *Graphic Masterworks of the Inuit: Kenojuak,* released the same year. Two hundred and seventy-five prints were editioned, and sixteen impressions were pulled bearing the Roman numerals I to XVI; all were individually signed by the artist.

Kenojuak *Myself and I*

Theo Waddington Inc., Folio of Lithographs, 1980

Following their commission and publication of the first lithographs by Kenojuak Ashevak in 1979 (Dorset 1980), Theo Waddington Inc. released a second collection of lithographs in 1980. This portfolio, individually signed by the artist, consists of six images, printed in an edition of fifty.

Kenojuak *Spirit of the Owl*

Kenojuak *Owl at the Centre*

Kenojuak *Colourful Sentinel*

Kenojuak *Days on the Coast*

Kenojuak *Guardian Owl*

Kenojuak *The World Around Me*

National Museums of Canada, 1979

This National Museums of Canada, 1979, commission and those following are works which were commissioned and released in earlier years. They have not been documented in previous annuals and are presented here to complete the record of works commissioned in Cape Dorset to date. National Museums of Canada, Ottawa, commissioned two works by Kananginak Pootoogook, which were released in 1979. These prints were in an edition of one hundred.

Kananginak *Sitgarriat I*

Kananginak *Sitgarriat II*

Skate Canada, World's '78

The Skate Canada competition, held in Ottawa, Canada, in 1978, commissioned a lithograph by Kananginak Pootoogook for distribution to the participants. The print, dated 1977, but released in 1978, was in an edition of seventy-five and signed by the artist.

Kananginak *Nanuk*

Waddington Galleries, Peter Pitseolak Folio, 1977

In 1977 Waddington Galleries released a portfolio of nine lithographs, after Peter Pitseolak, whose death occurred in 1973. The suite, editioned in 1974, numbered seventeen, and an additional five impressions were also pulled, bearing the Roman numerals I to V. The prints are signed by Aggeo, Peter Pitseolak's widow.

Peter Pitseolak *Two Girls Eating*

Peter Pitseolak *Dancer*

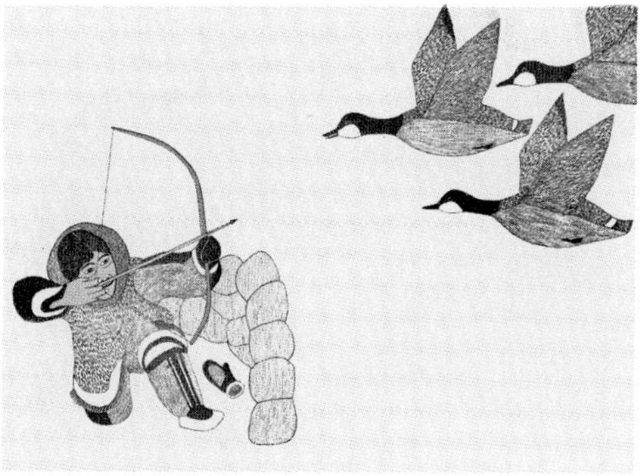

Peter Pitseolak *Nirliniaktuk*

Peter Pitseolak *Polar Bears*

Peter Pitseolak *Tatooed Woman*

Peter Pitseolak *A Feast of Char*

Peter Pitseolak *Foxes and Hares*

Peter Pitseolak *Tanisiqtut*

Peter Pitseolak *Summer Journey*

United Nations Conference, 1976

The Habitat United Nations Conference on Human Settlements, held in Vancouver, Canada, in 1976, commissioned a suite of four prints by individual artists. All four were in editions of fifty, and were released in 1976.

Jamasie *Animals and Men*

Kananginak *Winter Camp at Ikerasak*

Lucy *Thoughts of My Childhood*

Pudlo *Timiat Nunamiut*

Etching Portfolios

The third portfolio of copper etchings, released in 1981, illustrates the growing capabilities of three women who are primarily responsible for the success of the etching program in Cape Dorset. Auksuali Ottochie, Kooyoo Simiga and Udluriak Pudlat, as a team of printers, have allowed several artists the opportunity to see their work come to fruition.

In 1978 an etcher from Toronto was engaged temporarily by the Co-operative to assist with the renewed etching program. Since that time, and in spite of several lapses in continuity, all three women have remained with the program. In 1979 a second etcher from southern Canada assisted in further development of the project. At present, however, the women work alone under the supervision of resident lithographer, Wallace Brannen.

Continuing experimentation by the three technicians further enhances the potential of this ongoing project. Artists, aware of the expertise available to them, are showing an interest in expanding tonal and textural expressions in their work. Although content to work through the initial stages of drawing on a plate, few artists have expressed an interest or desire to involve themselves in the technical aspects of grounds, aquatinting, acid baths, or the monotony of editioning a run of prints.

All three women are married to men who are closely associated with printmaking. Timothy Ottochie cuts designs into the stone blocks used for printing; Ningoochiak Pudlat is a printer in the stonecut studio; and Simiga, occasionally a printer, is primarily a packer of the stone sculpture which is exported to southern markets.

The previous portfolios of etchings – the first, a selection of representative works by six individual artists; the second, dedicated to the works of Pitseolak Ashoona – are followed here by a selection of designs by three artists, Kenojuak Ashevak, Aoudla Pudlat and his stepfather, Oshoochiak Pudlat. The work of Kenojuak and Aoudla is well-known to collectors. Oshoochiak, however, a man in his early 70's, only began to draw in 1980. His work, published here for the first time, illustrates a unique approach both in subject matter and textural design.

Terry Ryan

Portfolio 111

Aoudla Pudlat *Tundra Bird*

Aoudla Pudlat *Surprised While Seal Hunting*

Kenojuak Ashevak *Owl and Bears*

Kenojuak Ashevak *Summer Sky*

Oshoochiak Pudlat *The Shaman's Family*

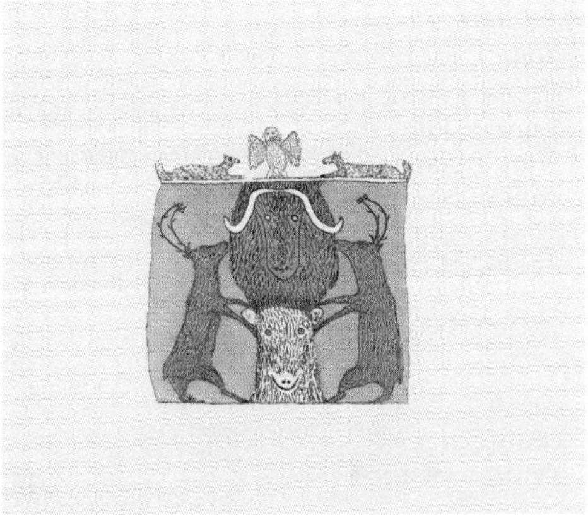

Oshoochiak Pudlat *Our Animals Together*

List of Graphics

All sizes are given in centimetres, height before width. Paper size will vary from print to print.

Cape Dorset Annual Graphics Collection 1981

Aoudla Pudlat

1
Imaginary Bird
Stonecut and Stencil
Printer: Kabubuwa Katsiya
46 x 60.5
Black, brown, green, orange, purple

Echalook Goo

2
Inuit
Stonecut and Stencil
Printer: Eegyvudluk Pootoogook
45.5 x 60.5
Black, blue, brown, green, purple, red

Eegyvudluk

3
Angako Quviasuktu
Stonecut and Stencil
Printer: Saggiaktok Saggiaktok
56 x 76.5
Editioned in 1978
Black, brown, green, yellow

4
Ishuk
Stonecut and Stencil
Printer: Kabubuwa Katsiya
61 x 80.5
Editioned in 1978
Black, brown, green, red, yellow

Eliyakota

5
Flying in a Strong Wind
Stonecut and Stencil
Printer: Eegyvudluk Pootoogook
62.5 x 86
Black, brown, green, yellow

Etidloie

6
Marching Birds
Stonecut and Stencil
Printer: Saggiaktok Saggiaktok
59.2 x 85.5
Black, brown, yellow

7
Birds Among Willow Trees
Stonecut and Stencil
Printer: Timothy Ottochie
63.2 x 86
Black, blue, brown, grey, orange, purple, yellow

Haunak

8
First Goose of the Spring
Stonecut and Stencil
Printer: Saggiaktok Saggiaktok
64 x 63
Black, brown, green

9
Owl Attacking Prey
Stonecut and Stencil
Printer: Pee Mikkigak
63 x 86
Black, blue, brown

Jamasie

10
Sailing in a Beautiful Boat
Stonecut and Stencil
Printer: Timothy Ottochie
63 x 86
Black, blue, purple

Kakulu

11
Disturbed While Nesting
Stonecut and Stencil
Printer: Saggiaktok Saggiaktok
63.5 x 86.5
Black, pink, yellow

12
The Clothing I Make
Stonecut and Stencil
Printer: Timothy Ottochie
62.5 x 86.5
Black, blue, orange, purple, yellow

13
Family Members Return
Stonecut and Stencil
Printer: Timothy Ottochie
63.5 x 86.5
Black, blue, green, grey, red

Kananginak

14
A Good Day Fishing
Stonecut and Stencil
Printer: Pee Mikkigak
63 x 86
Black, blue, brown, green, grey, orange, yellow

15
I Was at the Rudder
Stonecut and Stencil
Printer: Saggiaktok Saggiaktok
62.5 x 86
Black, blue, brown, grey, orange, yellow

16
Tiriak Battles the Owl
Stonecut and Stencil
Printer: Pee Mikkigak
62.5 x 86.5
Black, grey, orange, yellow

17
The Raven and the Goose
Stonecut and Stencil
Printer: Saggiaktok Saggiaktok
63 x 86
Black, brown, green, orange, yellow

18
Sharks Near Our Boats
Stonecut and Stencil
Printer: Eegyvudluk Pootoogook
63 x 86.5
Black, brown, grey, yellow

19
Feeding on Flies
Stonecut and Stencil
Printer: Saggiaktok Saggiaktok
63 x 77
Black, brown, red, yellow

20
Mother Protects Her Young
Stonecut and Stencil
Printer: Saggiaktok Saggiaktok
63 x 86
Black, brown, grey, red, yellow

21
Dance of Love
Stonecut and Stencil
Printer: Saggiaktok Saggiaktok
61.5 x 75.5
Black, brown, grey, orange, red, yellow

22
Sijariaq
Stonecut and Stencil
Printer: Saggiaktok Saggiaktok
62.5 x 72.5
Black, brown, orange, red, yellow

Keeleemeeoomee

23
Some Feed, Some Watch
Stonecut and Stencil
Printer: Timothy Ottochie
60 x 75.5
Black, brown, grey, orange

24
We Follow the Caribou
Stonecut and Stencil
Printer: Timothy Ottochie
62.5 x 77
Black, brown, grey, red

25
Two Birds Walking
Stonecut and Stencil
Printer: Timothy Ottochie
45.5 x 60
Black, green, orange

Kenojuak

26
Watching My Animals
Stonecut and Stencil
Printer: Saggiaktok Saggiaktok
63.2 x 86.5
Black, brown, green, orange, red, yellow

27
The Owl Rules the Summertime
Stonecut and Stencil
Printer: Ningoochiak Pudlat
46 x 60
Black, blue, brown, green, purple, yellow

Kingmeata

28
Talelayou and Fish
Stonecut and Stencil
Printer: Pee Mikkigak
63.5 x 86
Black, blue, green, orange, yellow

29
Joyful Young Man
Stonecut and Stencil
Printer: Timothy Ottochie
59.5 x 81
Black, blue, purple red, yellow

30
Hawk and Bear Converse
Stonecut
Printer: Saggiaktok Saggiaktok
49.5 x 61
Black, grey

31
Large Bird in Sunlight
Stonecut and Stencil
Printer: Timothy Ottochie
60.5 x 69.5
Editioned in 1980
Brown, black, orange, yellow

32
Talelayou and Spirit Fish
Stonecut and Stencil
Printer: Saggiaktok Saggiaktok
61 x 86.5
Black, blue, brown, green, orange, yellow

33
Bird Sentinel
Stonecut and Stencil
Printer: Timothy Ottochie
61 x 86.5
Black, green, orange,

yellow

Kudjuakjuk

34
Birds Eat Mosquitoes
Stonecut and Stencil
Printer: Timothy Ottochie
61.5 x 75
Blue, brown, green, yellow

35
Seagulls on Spring Ice
Stonecut and Stencil
Printer: Timothy Ottochie
63.3 x 86.5
Black, blue, grey

Lucy

36
Talelayou, Giant Birds and Men
Stonecut and Stencil
Printer: Timothy Ottochie
60.5 x 85
Black, blue, brown, green, grey, orange, yellow

37
Attacked by a Giant Bird
Stonecut and Stencil
Printer: Timothy Ottochie
63.5 x 86.5
Black, blue, brown, green

38
The Owl Looks On
Stonecut and Stencil
Printer: Timothy Ottochie
63 x 70
Black, brown, green, orange.

39
A Bird Steals a Fisherman
Stonecut and Stencil
Printer: Timothy Ottochie
59.5 x 71.5
Black, blue, brown, yellow

40
Birds Fishing
Stonecut and Stencil
Printer: Timothy Ottochie
62.5 x 86.5
Black, green, red, yellow

41
Trio of Owls
Stonecut and Stencil
Printer: Saggiaktok Saggiaktok
60 x 72
Black, grey, orange, yellow

42
Catching Moulting Birds
Stonecut and Stencil
Printer: Timothy Ottochie
30 x 41.5
Black, brown, grey, yellow

Napachie

43
A Tent at the Fish Lake
Stonecut and Stencil
Printer: Timothy Ottochie
50.5 x 71.5
Black, brown, green yellow

Pitseolak

44
The Eggs Begin to Crack
Stonecut and Stencil
Printer: Kabubuwa Katsiya
51 x 71
Black, blue, green, red, yellow

45
Dogs Eat the Seal
Stonecut and Stencil
Printer: Timothy Ottochie
42 x 48.5
Brown, green, grey, purple, red

46
Bird Escapes My Stone
Stonecut and Stencil
Printer: Saggiaktok Saggiaktok
50.5 x 63
Black, blue, brown orange, yellow

Pudlo

47
Ship of the Loon
Stonecut and Stencil
Printer: Timothy Ottochie
62.5 x 86.5
Black, blue, green, orange, yellow

48
We Travel With the Loon
Stonecut and Stencil
Printer: Kabubuwa Katsiya
63 x 86
Black, blue, orange, red, yellow

49
Loon Speaks to the Walrus
Stonecut and Stencil
Printer: Kabubuwa Katsiya
63 x 86.5
Black, brown, green, orange, purple, yellow

50
Ready to Fly
Stonecut and Stencil
Printer: Ningoochiak Pudlat
60 x 85.5
Black, brown, green, grey, yellow

51
The Loon and the Fish Speak
Stonecut and Stencil
Printer: Kabubuwa Katsiya
62.5 x 86
Black, blue, brown, orange, purple, yellow

52
Ptarmigan in Sunlight
Stonecut and Stencil
Printer: Kabubuwa Katsiya
62.5 x 87
Black, orange, yellow

Each of the prints, in an edition of 50 and printed on mulberry paper, is individually signed by the artist. In addition 5 proofs have been pulled. The printer is indicated by the personal chop which appears on each print.

Lithographs Dorset 1981

Aoudla Pudlat

L1
Colourful Bird and Fish
Printer: Aoudla Pudlat
Paper: Arches Buff
72 x 56.5
Black, blue, brown, orange, pink

L2
Bird with Large Fish
Printer: Aoudla Pudlat
Paper: Arches Buff
56.5 x 72.5
Black, brown, orange, purple

Etidloie

L3
Aircraft Becoming Sea Animals
Printer: Pitseolak Niviaqsi
Paper: Rives B.F.K.
51 x 66.5
Black, blue, brown, green, grey, pink, purple, yellow

Haunak

L4
Geese in Spring
Printer: Aoudla Pudlat
Paper: Rives B.F.K.
51 x 66.5
Black, green, grey

Jamasie

L5
Fish Weir
Printer: Aoudla Pudlat
Paper: Rives B.F.K.
51 x 66.5
Brown, green, grey, purple

L6
Walrus and Friend
Printer: Aoudla Pudlat
Paper: Rives B.F.K.
44 x 38
Black, blue, green, red

Kakulu

L7
The Animals Come Together
Printer: Pitseolak Niviaqsi
Paper: Rives B.F.K.
56.5 x 76.5
Black, blue, grey, yellow

Kananginak

L8
Nirlik
Printer: Aoudla Pudlat
Paper: Rives B.F.K.
43.5 x 60.5
Black, brown, yellow

L9
Kopinawa
Printer: Pitseolak Niviaqsi
Paper: Arches Buff
33 x 38
Black, brown, green, white

L10
Swan
Printer: Aoudla Pudlat
Paper: Rives B.F.K.
43.5 x 34.5
Black, yellow

L11
Metik
Printer: Pitseolak Niviaqsi
Paper: Arches Buff
37 x 47
Black, brown

L12
Tern Attacking Eider Duck
Printer: Pitseolak Niviaqsi
Paper: Rives B.F.K.
38 x 47
Black, blue

L13
Gulls at Sunrise
Printer: Aoudla Pudlat
Paper: Rives B.F.K.
46 x 59.5
Black, blue, pink

L14
Going into the Tent
Printer: Aoudla Pudlat

Paper: Arches Buff
38 x 48.5
Black, orange, purple

L15
Building a Kayakkovik
Printer: Pitseolak Niviaqsi
Paper: Arches Buff
38 x 48.5
Black, orange, purple

L16
Hunter at Aglu
Printer: Aoudla Pudlat
Paper: Arches Buff
35 x 43.5
Black, blue, yellow

Lucy

L17
Young Owl in Sunlight
Printer: Pitseolak Niviaqsi
Paper: Rives B.F.K.
43.5 x 58
Black, blue, green, orange, yellow

L18
Boy at Home Alone
Printer: Pitseolak Niviaqsi
Paper: Rives B.F.K.
52 x 66.5
Black, blue, brown, green, orange, pink, yellow

Mayoreak

L19
Putting Up the Tent
Printer: Pitseolak Niviaqsi
Paper: Arches Buff
54.5 x 65.5
Black, brown, green, white, yellow

L20
Cleaning Fish
Printer: Aoudla Pudlat
Paper: Rives B.F.K.
55.5 x 71
Brown, orange, yellow

Pauta

L21
Mother with Son
Printer: Pitseolak Niviaqsi
Paper: German etching
56.5 x 39.5
Black, orange, red, yellow

Pitaloosie

L22
Great Owl
Printer: Aoudla Pudlat
Paper: Nacre
56.5 x 76.5
Black, blue, green, pink

L23
The Young Owl
Printer: Pitseolak Niviaqsi
Paper: Arches Buff
51 x 66
Black, yellow

L24
Dancing Ptarmigan
Printer: Pitseolak Niviaqsi
Paper: Arches Buff
51 x 66
Black, blue, yellow

L25
Mother
Printer: Pitseolak Niviaqsi
Paper: Rives B.F.K.
66.5 x 50.5
Black, brown, green, purple, yellow

Pudlo

L26
Muskox and Friends
Printer: Aoudla Pudlat
Paper: Arches Buff
56.5 x 76.5
Black, blue, green, red

L27
Large Loon and Landscape
Printer: Pitseolak Niviaqsi
Paper: Arches Buff
56.5 x 76.5
Black, blue, brown, green, orange, red, yellow

L28
Boat and Airplane
Printer: Pitseolak Niviaqsi
Paper: Arches Buff
57 x 76
Black, blue, green, orange, purple, red, yellow

L29
Harpoons and Seals
Printer: Aoudla Pudlat
Paper: Rives B.F.K.
50.5 x 66.5
Black, blue, brown

All of the lithographs are in editions of 50. Each is signed by the artist. In addition, 3 trial proofs and 4 publishers' proofs have been pulled. Printers are indicated by the personal chop which appears on each print.

Commissioned Works

The Mintmark Press, 1981

Kenojuak

Myself and I
Hand-coloured Etching
Printed: black, green
Hand-coloured: blue, yellow
Plate: 20 x 25
Paper: 28.5 x 33.5

Theo Waddington Inc., Folio of Lithographs, 1980

Kenojuak

Spirit of the Owl
Lithograph
Printer: Pitseolak Niviaqsi
Paper: German etching
56.5 x 79
Black, blue, brown, green, orange, red, yellow

Owl at the Centre
Lithograph
Printer: Aoudla Pudlat
Paper: German etching
56.5 x 79
Black, blue, brown, green, orange, purple, yellow

Colourful Sentinel
Lithograph
Printer: Aoudla Pudlat
Paper: German etching
56.5 x 78.5
Black, brown, orange, yellow

Days on the Coast
Lithograph
Printer: Aoudla Pudlat
Paper: German etching
56.5 x 79
Black, brown, grey, yellow

Guardian Owl
Lithograph
Printer: Aoudla Pudlat
Paper: German etching
56.5 x 78.5
Black, brown, green, orange, purple, red, yellow

The World Around Me
Lithograph
Printer: Pitseolak Niviaqsi
Paper: German etching
56.5 x 79
Black, blue, brown, green, purple, red, yellow

National Museums of Canada, 1979

Kananginak Pootoogook

Sitgarriat I
Engraving
Plate: 25 x 30
Paper: 39 x 45.5
Paper: German etching
Brown

Sitgarriat II
Etching and Aquatint
Plate: 25 x 30
Paper: 39.5 x 51
Paper: German etching
Brown, green

Skate Canada, World's '78

Kananginak Pootoogook

Nanuk
Lithograph
Printer: Pitseolak Niviaqsi
Paper: German etching
25.5 x 30
Black, yellow

Waddington Galleries, Peter Pitseolak Folio, 1977

Peter Pitseolak

Two Girls Eating
Lithograph
Printer: Pee Mikkigak
Paper: Rives B.F.K.
51 x 66
Black, blue, brown, green, orange, red

Dancer
Lithograph
Printer: Pee Mikkigak
Paper: Arches Buff
56.5 x 74
Black, brown, green, orange, purple, red

Nirliniaktuk
Lithograph
Printer: Pitseolak Niviaqsi
Paper: Rives B.F.K.
49 x 64
Black, brown, grey, orange, pink

Polar Bears
Lithograph
Printer: Aoudla Pudlat
Paper: Rives B.F.K.
56.5 x 75.5
Black, red, yellow

Tatooed Woman
Lithograph
Printer: Pee Mikkigak
Paper: Rives B.F.K.
66 x 50.5
Black, blue, brown, green, orange, red, yellow

A Feast of Char
Lithograph

Printer: Pitseolak Niviaqsi
Paper: Rives B.F.K.
61 x 46
Black, brown, green, red

Foxes and Hares
Lithograph
Printer: Pee Mikkigak
Paper: Arches Buff
50 x 65
Black, grey

Tanisiqtut
Lithograph
Printer: Pitseolak Niviaqsi
Paper: Rives B.F.K.
50.5 x 66
Black, brown, green, orange, purple, red, yellow

Summer Journey
Lithograph
Printer: Pitseolak Niviaqsi
Paper: Rives B.F.K.
50.5 x 66
Black, brown, orange, yellow

United Nations Conference, 1976

Jamasie

Animals and Men
Stonecut
Printer: Saggiaktok Saggiaktok
Paper: Mulberry
63 x 86.5
Green, red

Kananginak

Winter Camp at Ikerasak
Stonecut and Stencil
Printer: Timothy Ottochie
Paper: Mulberry
61.5 x 85
Black, blue, brown

Lucy

Thoughts of My Childhood
Stonecut and Stencil
Printer: Timothy Ottochie
Paper: Mulberry
61.5 x 85
Green, grey, red, yellow

Pudlo

Timiat Nunamiut
Hand-coloured lithograph
Printer: Pitseolak Niviaqsi
Paper: Arches
60 x 91
Black, blue, orange, purple, yellow

Etchings Dorset 1981

Portfolio III

Aoudla Pudlat

Tundra Bird
Etching and Aquatint
Plate: 25 x 30
Paper: 38 x 43
Brown, green

Surprised While Seal Hunting
Hand-coloured Etching and Aquatint
Plate: 25 x 30
Paper: 38 x 43
Printed: black
Hand-coloured: brown, yellow

Kenojuak Ashevak

Owl and Bears
Etching and Aquatint
Plate: 20 x 25
Paper: 38 x 45.5
Black, green, purple, yellow

Summer Sky
Etching and Aquatint
Plate: 20 x 25
Paper: 38 x 46
Black, blue, green, orange, purple, red.

Oshoochiak Pudlat

The Shaman's Family
Hand-coloured Etching and Aquatint
Plate: 25 x 37
Paper: 38 x 38
Printed: Black, grey
Hand-coloured: brown

Our Animals Together
Etching and Aquatint
Plate 25 x 24.5
Paper: 38 x 38
Blue, brown, yellow

All of the etchings are in editions of 25, and are printed on German etching paper.

Errata

Due to production errors, print numbers have been reversed in four instances, as follows:

Dorset 75

Page 24 – reverse numbers 23 & 24
Page 41 – reverse numbers 49 & 50
Page 53 – reverse numbers 69 & 70

Dorset 80

Pages 68 & 69 – reverse numbers 53 & 54

In the introduction to *Dorset 80* Virginia Watt's reference to the print "Young Mother with Owls" refers to the print "Arctic Madonna" by Pitaloosie.

Edition Numbers
It should be noted that edition numbers, e.g. 1/50, do not necessarily indicate the order in which a series of images were pulled. Numbers are allocated at the time of print examination and signing.

The prints have been selected by the Canadian Eskimo Arts Council, an advisory body which recommends policy to the Minister of Northern Affairs and helps orderly development of Eskimo art markets. The committee is drawn from known and established members of the Canadian art community.